Wired World: A Beginner's Guide to Embedded Electronic Interfaces

Omar Hiari

Preface

Embedded systems often pose a formidable hurdle for newcomers due to the intricate world of electronics they delve into. Navigating through the labyrinth of circuits and technical jargon is an unavoidable part of configuring peripherals in a controller. At times, even deciphering a schematic becomes a necessity to comprehend the external circuitry. This challenge can be particularly daunting for those aspiring to enter the embedded realm without a strong hardware background. Basic electronics knowledge becomes a prerequisite to grasp these foundational concepts.

Coming from a hardware-oriented background myself, I've noticed a recurring issue in many educational resources: they often attempt to cover an extensive range of electronics topics for beginners. While this comprehensive approach is crucial for those aspiring to design intricate circuits, it may deter individuals primarily interested in embedded software.

I firmly believe that there exists a simple critical subset of electronic circuits simple enough for beginners in the embedded field to grasp. This subset may well suffice for embedded software engineers to embark on a successful career journey. The silver lining is that embedded electronics generally prioritize simplicity, as complexity typically leads to increased costs and power consumption.

Motivated by this realization, I have undertaken the task of crafting this book to bridge the existing gap. My goal is to equip readers with a comprehensive understanding of the most prevalent electronic circuitry they will encounter at the outset of their journey. My hope is that this resource will empower you and prove invaluable in your endeavors within the embedded world.

Thank you for joining me on this educational adventure.

-Omar

Twitter/X: apollolabsbin
Blog: https://apollolabsblog.hashnode.dev/

Please log any book issues on the book repo here:
https://github.com/apollolabsdev/wiredworldbook

Table of Contents

0. Background

This chapter covers some essentials you might find useful to reference throughout the book.

0.1 Table of Acronyms

ACRONYM	DEFINITION
OC	Open Circuit
CC	Closed Circuit
DC	Direct Current
AC	Alrentating Current
GPIO	General Purpose Input Output
ADC	Analog to Digital Converter
DAC	Digital to Analog Converter
PWM	Pulse Width Modulation
BJT	Bipolar Junction Transistor
FET	Field Effect Transistor
Hi-Z	High Impedance
LED	Light Emitting Diode
Op-Amp	Operational Amplifier
dB	Decibels

ACRONYM	DEFINITION
LPF	Low Pass Filter
HPF	High Pass Filter
BPF	Band Pass Filter
UART	Universal Asyncronous Recieve Transmit
SPI	Serial Peripheral Interface
USB	Universal Serial Bus
CAN	Controller Area Network
IC	Integrated Circuit
i/p	Input
o/p	Output
uC	Microcontroller
rpm	Revolution Per Minute
CW	Clockwise
CCW	Counter Clockwise

0.2 Simulation Tools

Interactivity can be of huge help when learning. However, learning hardware and electronics sometimes suffers the challenge of needing special equipment. Equipment that depending now what you need done, might be expensive. While this might have been an impediment in the past, it has become better in many areas. This is because of simulation tools that emerged that allows one to simulate hardware using software. In that context all you do not need more than just a computer. The following is a list of some useful tools:

NAME	COST	LEVEL
Autodesk Tinkercad	Free	Beginner
Wokwi	Free	Beginner Intermediate
Proteus	Paid	Intermediate Advanced
Virtual Breadboard	Hybrid	Beginner
Picsimlab	Free	Beginner

Note that these tools are ones that are embedded systems oriented. Meaning that you can integrate some form of microcontroller hardware and code it. There are other tools out there specific to circuit simulation only, however I did not include any of those. Those are more advanced tools better visited as one gets more comfortable with electronics and hardware design. The most popular being SPICE-based simulation software.

In this book I will be using Tinkercad for interactive examples and exercises. These examples will be highlighted with the following icon: 🎮. There will be a link to each example that you can tinker with on your own. Tinkercad is fairly intuitive to use, however, if you are new to or not comfortable with Tinkercad, there are some beginner tutorials to help get you started.

! For editions of the book that do not support hyperlinks please refer to the appendix for the raw links.

0.3 Open vs Closed Circuit

In this book there is going to be a lot of reference to open and closed circuits. As such, we must highlight the difference between an *open circuit* and *closed circuit*. Figure 1 demonstrates two switch operated circuits. The circuit on the left-hand-side with the switch closed demonstrates a *closed circuit* where current is flowing through the circuit without interruption. The circuit on the right-hand-side with the switch open demonstrates an *open circuit* where current flow is interrupted. Essentially, in a *closed circuit* there is current flow through a load/resistance whereas in an *open circuit* there isn't.

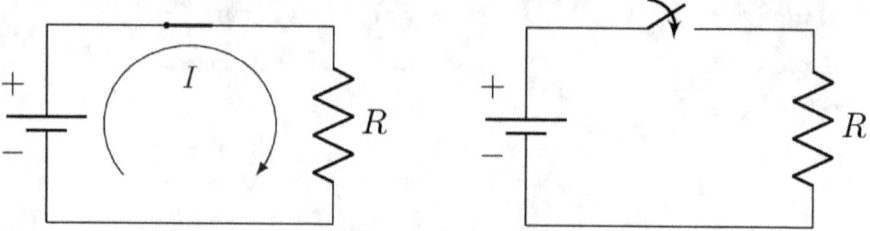

Figure 1. Closed and Open Circuit Comparision

0.4 Power Symbols

Going forward in this book we are going to omit the explicit drawing of the battery and its negative and positive terminals. Instead we are going to replace the battery with symbols. Figure 2 shows those symbols. Both the left-hand-side and right-hand-side circuits are exactly the same operationally. However, we're going to replace the battery positive terminal with a symbol of an arrow pointing upwards. V_{cc} is notation indicating the battery voltage. The voltage indicated at the top of the arrow is also referred to as the **upper rail**. On the other hand, the negative battery terminal will be replaced with a symbol that is a horizontal line with three downward-pointing lines parallel to it. It's going to be referred to as ground or the **lower rail**. Why are we doing this? These symbols are more visually appealing, more convenient to use, and eliminate the need for drawing a lot of wires. There's no need to draw a battery or power source all the time, only indicate its voltage output.

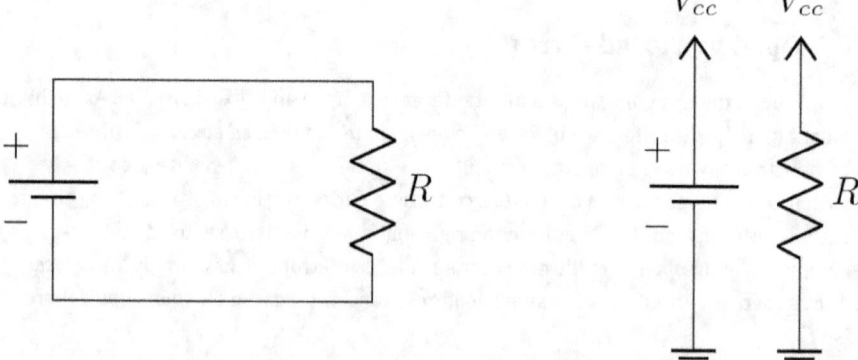

Figure 2. Alternative Symbols for Positive and Negative Battery Terminals

0.5 DC vs AC Signals

The most important thing we need to know is that direct current (DC) signals have a constant voltage and/or current and a frequency of 0 Hz. On the other hand, alternating current (AC) signals vary their voltage and/or current levels and have frequencies more than 0 Hz (Ex. A sine wave). Signals could also be a combination of DC, AC, and even AC with different frequencies.

0.6 Simulation Examples

DC vs AC Signals: This simulation will help you explore the difference between AC and DC signals. The simulation also shows you the resulting signal when two types of signals are added. In the example, there are four oscilloscope circuits each representing a different case. Also function generators are used to generate the different signals. You can think of a function generator as a power source (Ex. like a battery) that can generate a variety of waveforms like sine waves and others. All the outputs are connected to oscilloscopes. An oscilloscope is a tool that allows us to view output waveforms. Note that on the y-axis of the oscilloscope expresses the voltage value while the x-axis expresses time. Beware that although the signals might look similar sometimes, their voltage scale in the y-axis might be different.

Try the following:

- If you click on a function generator, note a DC offset value. If you change it, what do you notice? Note that if you change it for any of the AC examples, you see an effect similar to the AC+DC example.

- In any of the AC examples, change the frequency of the input signals and see how the output changes.

1. Introduction

Embedded systems enable the connection (or interface) of the external world to the digital world, allowing us to both sense the environment and control events happening in it. Consequently, microcontrollers provide a vehicle in which that interaction with the external world is enabled. A microcontroller is an orchestrator that sits at the center of any embedded system and interfaces to the outer world through its peripherals. Peripherals are the entities that connect directly to physical pins on a microcontroller interfacing with the outer world through electrical signals. These peripherals include GPIO, ADCs, DACs, Timers and Counters, PWM, and Serial Communication interfaces among others. Additionally, each one of these peripherals have different electrical signal properties and considerations when interfacing with the outer world. These considerations are typically addressed through adding special electronic circuits to condition or transform the external signals to become either compatible with inputs to the microcontroller or outputs the microcontroller is controlling. Sometimes, the conditioning is offered as part of the microcontroller internal circuits and can be configured as such.

Figure 3 offers a representation of the levels of interfaces that exist around a microcontroller. The outer most level representing the outer world is typically analog in nature and can either consist of different device types or even introduce different effects. Devices and effects can come in different forms and values. As such, the next level of interface electronics purpose can be viewed as two-fold:

1. Protect the microcontroller interface against external outer world effects. For example, protect from **noise** or incorrect connections/configurations.

2. Adapt external device interfaces to become compatible with the microcontroller interface. For example, adjust input voltage levels or provide more drive current.

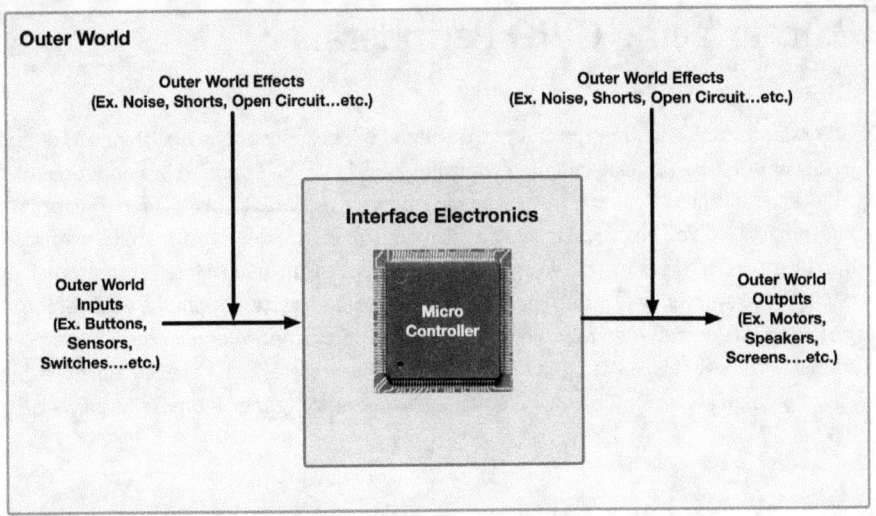

Figure 3. Microcontroller Interfaces to the Outer World

This e-book covers some of the most common considerations for the different peripherals and the interface electronics that go along with them. The concepts covered in this e-book are the most common that one would encounter in many systems. This book is meant to be as a reference for beginner developers interested in entering the embedded systems world but don't have an extensive (or any) background in electronics. As such, the concepts are explained in a manner and that only requires a basic understanding of electrical circuits.

The primary aim of this book is not to immerse readers in the intricacies of electronic circuit theory, but rather to assist them in recognizing familiar circuitry when encountered. Alternatively, if one comes across a controller setup, the goal is to provide a basic understanding of its significance. Ideally, the objective is to grasp the underlying purpose of a specific circuit or configuration and gain insight into its function.

Throughout this book, important terms commonly used will be in **_bold italics_**.

2. Interfacing GPIO Peripherals

GPIO stands for General Purpose Input/Output and it describes behavior of pins as digital inputs or outputs. Typically all microcontroller pins have the option to be configured to GPIO at a minimum. A digital signal can have two active states either as **high** (binary 1) or **low** (binary 0). At an electrical level at a microcontroller pin, these forms are transformed into electrical voltages measured against ground. As a result, a **low** state is represented by 0 Volts or alternatively Ground. On the other hand, a **high** state, adopts the value of the system/microcontroller voltage. For example, if the microcontroller runs on a power source/battery that is 3.3V then a **high** state would be represented by 3.3V. Also, note that most common system/microcontroller voltages that one might encounter nowadays is 3.3V or 5V.

2.1 Interfacing Digital Inputs

2.1.1 Input Change Detection

Changes on GPIO input pins can be configured for detection in two ways; either **polling** or **interrupts**. **Polling** is essentially the processor continuously checking if the input pin state has changed. This can be really wasteful of a processor's time as the change can be infrequent. **Interrupts** on the other hand allow changes (or events) in hardware (on the pin) to call software routines. This provides for a more efficient implementation and good use of a processor's time. Rather than having the processor waste time continuously checking a GPIO input for a change (Ex. a button press), the GPIO input can instead inform the processor of the change event. As such, input pins when configured as **interrupts** can be configured for two types of detection; either **edge detection** or **level detection**. These are also often referred to as **triggering mechanisms**.

Figure 4 demonstrates visually the different types of events that can be detected in an input digital signal. **Edge detection** involves identifying transitions in a digital signal. There are two types of edge detection, **rising edge detection** that marks a transition from a **low** state to a **high** state and **falling edge detection** that marks a shift from **high** state to **low** state. **Rising edge detection** is sometimes also referred to as **positive edge detection**, whereas **falling edge detection** is sometimes also referred to as **negative edge detection**. Note that commonly these events are marked with arrows on the signal.

Level detection focuses on recognizing when a signal reaches a specific predefined threshold or level to trigger a specific action or response. Similar to *edge detection*, an event is detected when this threshold is exceeded or falls below a certain value. There are also two types of *level detection*; *low-level detection* and *high-level detection*.

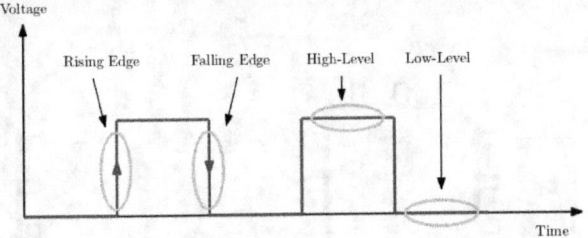

Figure 4. Types of events that can be detected on an input signal

It becomes obvious that having known voltage levels at input pins is crucial to avoid false-positive events.

2.1.3 Floating Pins

Consider the two circuits in Figure 5. These circuits show two ways one might connect a push button (or a switch) to a microcontroller pin to detect its state. Note that when the button is in the pressed state then this forms a *closed circuit* allowing current to flow and the voltage measured at the GPIO pin V_{pin} is either 0 V (*low* state) or 5 V (*high* state).

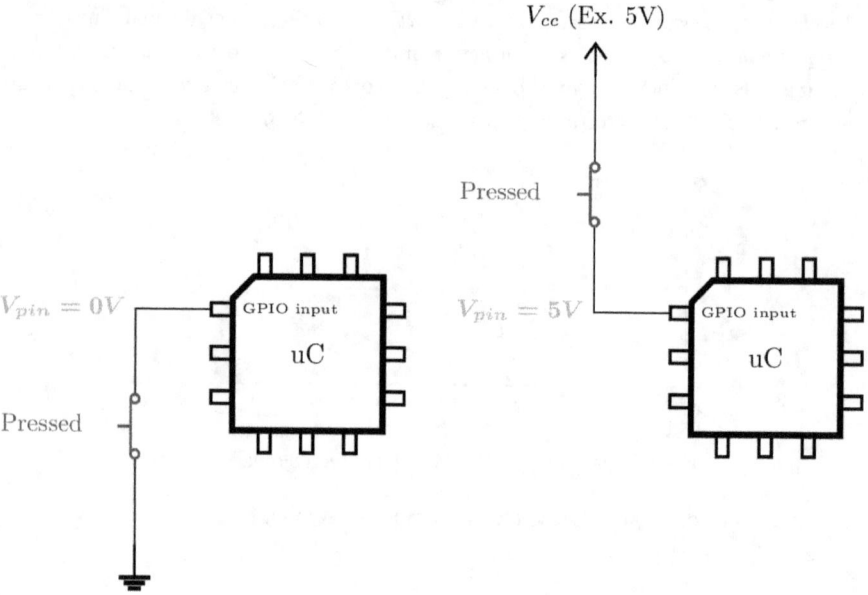

Figure 5. Observed Voltage at GPIO Inputs for Different Button Configurations when a Button is Pressed

Now observe Figure 6 when the button is unpressed. For both configurations an **open circuit** is formed preventing current from flowing and the pin is left as what is often referred to as **floating**. As a result, the voltage V_{pin} observed at the pin is unknown and could be any random voltage. Actually with enough surrounding **noise** one might observe a large enough voltage reading on the pin. Also, a common mistake many beginners make is assume that **floating** pins represent a **high** state.

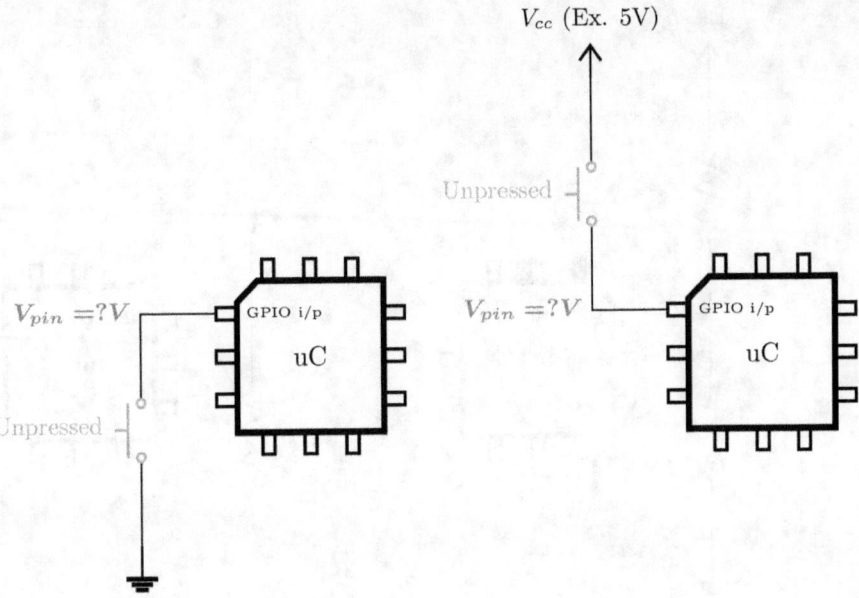

Figure 6. Observed Voltage at GPIO Inputs for Floating Button Configurations when a Button is Unpressed

Obviously this is a problem which needs solving. We cannot have an unknown voltage when the button is unpressed. This would potentially lead to many false positives depending on the environment. This is where **pull-ups** and **pull-downs** come in.

2.1.4 Pull Ups and Pull Downs

So our issue in the previous section was that the circuit is open when the buttons were unpressed and the pins were *floating*. This means that we need to somehow provide a known voltage (close the circuit) or default state when the button is unpressed. The solution can come in the form of introducing a simple resistor. This resistor would connect either to ground or V_{cc} depending on the button configuration. This is demonstrated in Figure 7 & Figure 8.

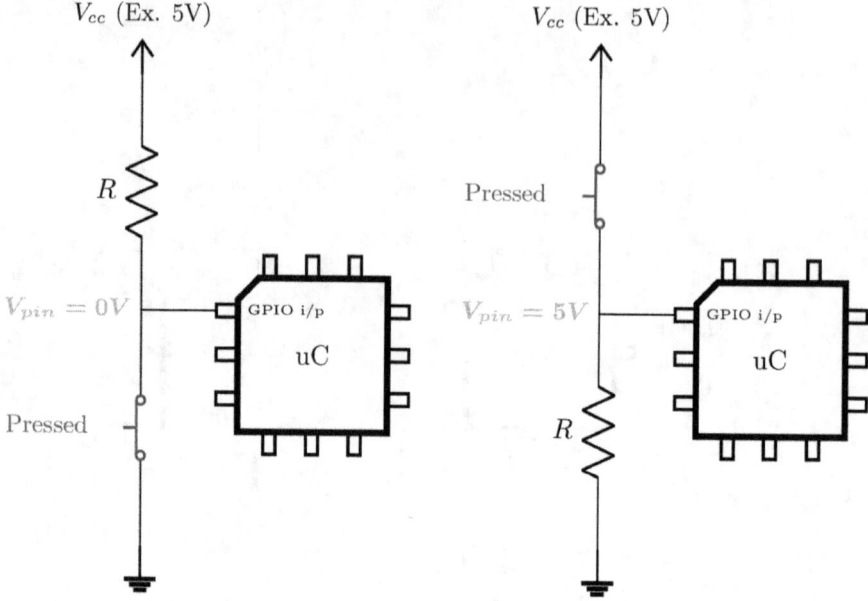

Figure 7. Observed Voltage at GPIO Inputs for Button Configurations with Pull Reistors when a Button is Pressed

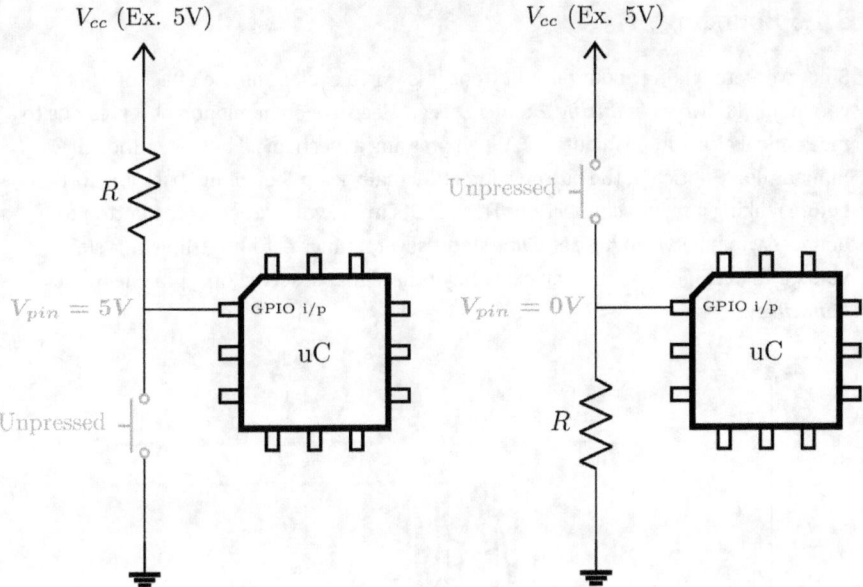

Figure 8. Observed Voltage at GPIO Inputs for Button Configurations with Pull Reistors when a Button is Unpressed

Note how now when the button is unpressed, we have a **closed circuit**. As a result, now a known voltage will appear at the pin that is close to either V_{cc} or ground when the button is unpressed. These resistors introduced are referred to as **pull-up** and **pull-down** resistors. They are named as such because a **pull-down** resistor pulls the voltage "down" to ground whilst a **pull-up** resistor pulls the voltage "up" to an **upper rail** or in this case V_{cc}.

Pull-up and **pull-down** resistors typically have values between $4.7k\Omega$ to $10k\Omega$. Additionally, many microcontrollers allow designers to add **pull-up** and **pull-down** resistors internally by software configuration. This relieves the designer of adding the resistors external to a microcontroller. Generally, for any pin, you should typically refer to a schematic to see what the external circuitry of a pin looks like. For example, if an external **pull-up** is placed, then you don't need to configure one internally.

2.1.5 Bouncing

Since we were talking about push buttons it is worth explaining the effect of what is commonly known as **bouncing**. **Bouncing** is an electrical phenomenon observed due to a mechanical effect in push buttons. When pressing a mechanical button, although we humans don't notice it, the button contacts actually keep "bouncing" back and forth before making a secure connection. This results in the voltage across the button to bounce/switch between **high** and **low** states several times before settling on a stable voltage value. Figure 9 shows an example of the signal observed on a pin due to the **bouncing** effect.

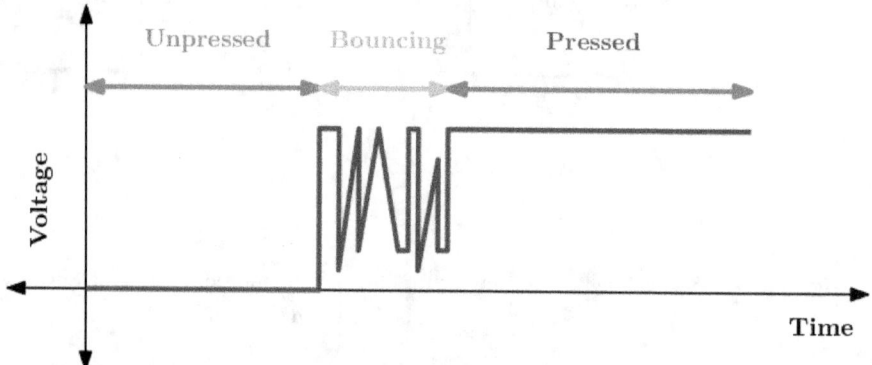

Figure 9. Effect of Bouncing on Press Button Signals

There are both hardware and software techniques to address bouncing. A simple solution in software is to take several consecutive readings of an input pin. If all the samples have the pressed state then a press is confirmed. The number of samples really depends on how fast your controller is but you can test by trial and error. For example if a press is detected, a condition could be that you see at least 5 consecutive pressed readings before you confirm a button press. Nevertheless, this might not be the most efficient approach. The good news is that there are many open source libraries out there that provide **debouncing** algorithms. Alternatively, if you want to get more intimate details, A Guide to Debouncing by Jack Ganssle is a recommended place to start.

2.2 Interfacing Digital Outputs

2.2.1 Transistors

Ahead of diving into interfacing digital outputs, it's essential to acquire some basic knowledge of transistors. Most transistors consist of three terminals and come in different types exerting different properties. However, the operating principles are more or less the same. Figure 10 shows two types of transistors. The transistor symbol on the left is known as the Field Effect Transistor (FET). A FET has three terminals; a **Gate** (G), a **Source** (S), and a **Drain** (D). The transistor symbol on the right is a Bipolar Junction Transistor (BJT). Similar to a FET, a BJT has three terminals though with different names; a **Base** (B), a **Collector** (C), and an **Emitter** (E). Many transistors out there are variants of these two types. What we are concerned here with is operation and they are similar conceptually.

The **gate** (or **base** in a BJT) is the terminal used to control the flow between the other two terminals. While there are different ways in which a transistor can operate, for now we are only interested in one; operating the transistor as a switch. Essentially, in switch operation, you can think of the transistor as a faucet where the **gate** (or **base**) is the handle and the **drain** (or **collector**) is where the water is being held to flow toward the **source** (or **emitter**). When operating as a switch, current is either allowed to flow completely between the **drain** (or **collector**) and the **source** (or **emitter**). Essentially opening the faucet all the way or closing the faucet all the way (cut off flow).

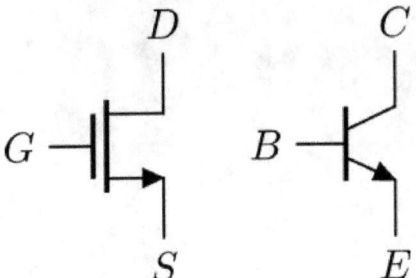

Figure 10. A Field Effect Transistor Symbol (left) and A Bipolar Junction Transistor (right)

2.2.2 High Impedance Outputs

One operating mode of GPIO outputs is known as **High impedance (or High-Z/Hi-Z)**. Impedance can be considered as another name for resistance. When a pin is **high impedance** it effectively means that it minimizes (ideally eliminates) current flow into a pin creating an **open circuit**. Consequently, if an output pin is configured as **Hi-Z** it can be considered effectively disconnected from the circuit it's connected to. In this state, the pin doesn't actively drive the pin **high** or **low**. It's like unplugging a wire: the pin doesn't actively send any signal, allowing external components to influence the voltage on that pin. This property is useful for scenarios where you want to prevent conflicts or interference when multiple devices are connected to the same wire/bus or when you want to use the pin as an input without affecting the rest of the circuit. As a matter of fact, pins when configured as inputs are by default in **Hi-Z** state.

2.2.3 Open Drain Outputs

An **open-drain** GPIO output pin is a microcontroller pin configured as an output that can only actively pull the pin **low**, but nothing else. Figure 11 shows how the internal circuit of the pin looks like in this configuration. We obviously can drive the pin **low** in this configuration. However, does this mean that the pin cannot be driven **high**? Actually, no, but the pin rather requires an external circuit such as a **pull-up** resistor to do so. This setup is useful for creating bidirectional communication lines and allowing multiple devices to share the same line without conflicts. It's like a valve that can drain current to ground but requires external help to refill back to a high level.

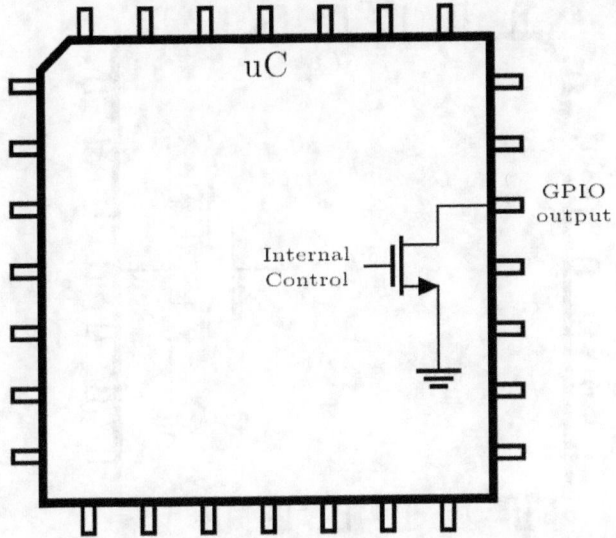

Figure 11. Internal Circuit of an Open Drain GPIO Output

Similar to input pins, it's worth noting that there are microcontrollers that allow configuring *open-drain* outputs with internal *pull-up* resistors.

2.2.4 Push-Pull Outputs

Push-pull is the most common output configuration found in microcontrollers. *Push-pull* GPIO outputs are types of pins on a microcontroller that can actively drive an output to both *high* and *low* states. A *push-pull* output consists internally of two transistors as shown in Figure 12. In the *low* output state, the bottom transistor is switched on "pulling" the pin to ground creating a *low* state. This is while the upper transistor is switched off (*open circuit*). In the *high* state, the lower transistor turns off creating an *open circuit* while the upper transistor turns on. As a result, the upper transistor "pushes" the pin to a voltage level close to the microcontroller supply voltage, creating a *high* state.

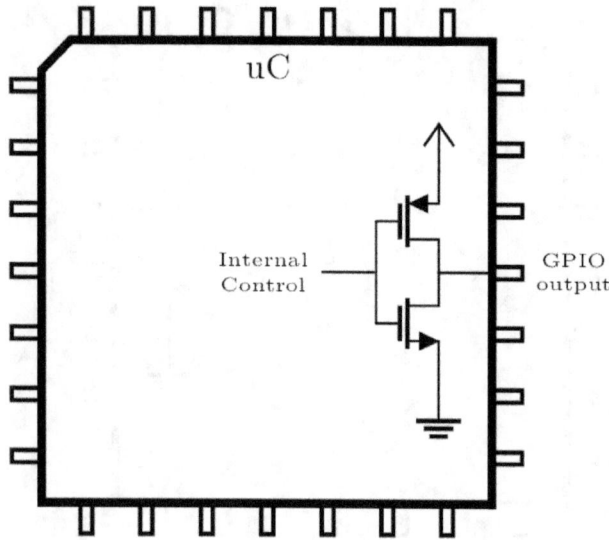

Figure 12. Internal Circuit of an Push Pull GPIO Output

A *push-pull* output allows the pin to actively source (when **high**) or sink (when **low**) current, making it suitable for driving various loads like LEDs, motors, or other external devices. However, a ***push-pull*** output isn't suitable for connecting multiple devices together on a single wire (Ex. a bus configuration), like the ***open-drain*** output. As such, a ***push-pull*** configuration is mostly suitable for interfaces with unidirectional control.

2.2.5 Drive Current

GPIO output pins are suitable for driving/controlling various loads, nevertheless, to a certain limit. Additionally, some controllers have additional drive current configurations that allow higher ***drive current*** to be sourced from some output pins. Though the current sourced can be configured to provide up to tens of milliamps maybe but not more. While this might be enough to drive low power devices, it is not enough for many others. In the end of the day, a microcontroller is not a power source, it's more or less an intelligent control device. So what if you have an external load you need to control that requires more than a few milliamps (Ex. a motor)? This is where external ***drivers*** come in.

2.3 Drivers

Embedded applications entail the control of devices/actuators like motors, solenoids, speakers, heating elements, and LEDs, among others. These devices could require a range of drive currents from a few milliamps up to a few Amperes. Consequently, in some cases a microcontroller output pin cannot provide or source enough current to "drive" or control such an actuator. ***Driver*** circuits offer a solution by allowing a microcontroller low current output to "drive" a higher current from a different, higher current, power source. This means that the microcontroller would be able to control power to an external device but does not have to source the power itself. This is enabled by devices like power transistors which operate similar to regular transistors but rather can handle higher currents.

In the following sections some of the common ***driver*** configurations are introduced. Please note that the idea from the following sections is to introduce the general form of driver configurations and is in no way comprehensive. This would enable the designer to at least identify different configurations and how to control them. As this book focuses on the beginner learner, the detailed design behind driver circuits is out of the scope of this book.

2.3.1 Driving an external load/device through a microcontroller output pin

This is one of the simplest form/configurations of a ***driver*** and the one we've been referencing thus far. This is a configuration good for low current applications. In this configuration the output pin is connected directly to the load/device we want to control. The Figure 13 shows two of the common examples found on almost every embedded development board. An LED driven by a GPIO output to exercise a Blinky example.

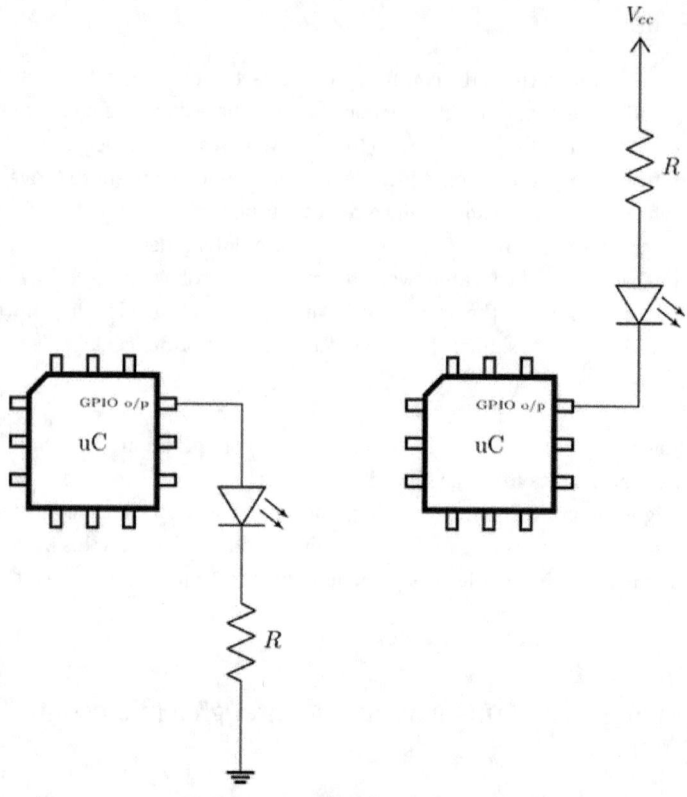

Figure 13. LEDs Driven by GPIO Output

Note that there are two possible configurations, active **high** and active **low**. The only difference is that one activates the LED when the output is **high** (left configuration) and the other when the output is **low** (right configuration). The microcontroller pin can be configured as a **push pull** output for both configurations. However, for the configuration on the right specifically, the pin can be configured as an **open drain** output. This type of **driver** is sufficient for the range of a few milliamps with the exact value of drive current defined in the microcontroller data sheet.

> *Note that between the two configurations in Figure 11 the direction of current differs.*
> *On the left hand side the current comes out from the pin towards ground. However, on*
> *the right hand side the current moves toward the GPIO pin to the internal ground of*
> *the microcontroller.*

2.3.2 Driving an external load/device through a transistor

Let's say we have different type of LED (Ex. higher intensity) that requires a few hundred milliamps. Or alternatively a device like a motor that requires a few Amps. In this case a (power) transistor and a separate power source/battery can be introduced to form the configuration in Figure 14.

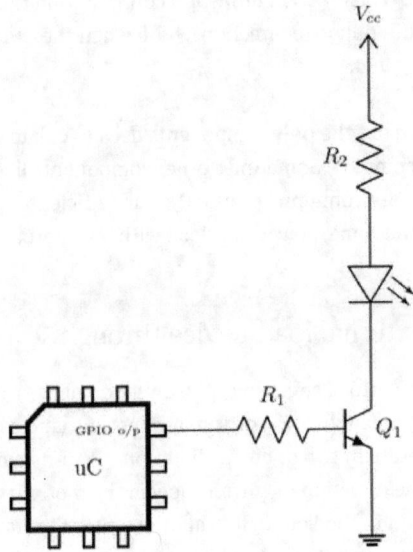

Figure 14. Driving and LED with an External Transistor

This type of driver circuit is similar to the **open drain** configuration we covered, albeit external. Though in this case the accurate term is **open collector** as the transistor used is a BJT transistor. This type of circuit is designed with the following considerations:

1. The transistor is configured to operate as a switch where it either allows current to flow between the source and the drain (and thus the LED) or stops it.

2. A GPIO output is used only to control or turn the transistor on and off. You can think of the GPIO output as the human finger pressing a button to make contact between the other two transistor terminals. In this configuration, the transistor typically does not require much, if any, current at the base to control it.

3. The LED current is sourced from a separate power source (V_{cc}) connected at the end of the drain terminal.

In addition to the above, the **push-pull** configuration discussed earlier can also be observed as an external **driver** circuit. Again, the difference being is that the transistors can handle higher currents and run off a separate source. The question is when is a **push-pull** driver configuration needed? The simple answer is that some application loads require continuous current flow in both directions. Note how in the **open drain** configuration when the transistor is off (Ex. GPIO output **low**) there is no current flowing in the load/LED (**open circuit**). With a **push-pull** configuration this is not the case. Instead current will flow, albeit in opposite directions, both when the GPIO output is driving the transistors with a **low** or **high** state.

Note that transistors are not the only components used to create driver circuits though probably are the most common. Sometimes other components like operational amplifiers are utilized. Transistors are common because they are efficient and cheap. The idea, nevertheless, remains the same, powering a load with an external source.

2.3.3 Driving an external load/device through several transistors

In applications that use motors, sometimes the direction of the rotation needs to be altered. For example, in an application driven by a motor that closes and opens a door, the motor would need to rotate in two opposite directions. It turns out that motor rotation direction depends on the direction of current flow in the motor itself. As such, using an **open-drain** configuration is good enough to allow a motor to rotate in one direction. Moreover, a **push-pull** configuration would allow current to continuously flow in opposite directions, which means there is no way to stop the motor from rotating. This is where configurations like the **H-bridge** shown in Figure 15 come in.

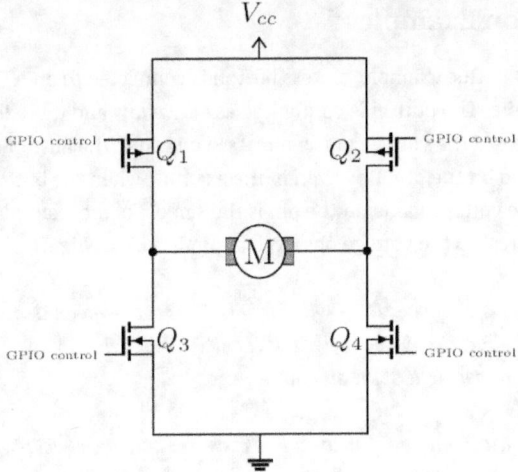

Figure 15. An H-Bridge Conifguration

Note how in the **H-bridge** circuit there are four transistors used to control current flow direction. If all transistors are off, then there is no current flow and the motor is stopped. If the upper left and lower right (Q1 and Q4) transistors are turned on (all others off) then the current flows in the motor from left to right. Alternatively, if the upper right and lower left (Q2 and Q3) transistors are turned on (all others off) then the current flows in the motor from right to left.

> *In real world implementations, common H-bridge motor driver devices used integrate the transistors in one package/device. As a result, it's unlikely that one would see four separate transistors hanging around in a schematic. It's still possible for one to implement using individual transistors but it's considered less efficient and takes more space.*

Given what we know thus far, using regular digital output to drive external loads might introduce a challenge. This is when we want to control the amount of current delivered to a device. In existing configurations, power to the external load is either turned fully on or fully off but nothing in between. However, in some applications the amount of current delivered to the load needs to be controlled. Examples include controlling motor speed and LED intensity. As such, trying to control the current delivered through GPIO outputs would require more software involvement. However, an alternative solution comes in the form of pulse width modulation (**PWM**) which is introduced in the next chapter.

2.4 Simulation Examples

👾 Floating Pins: In this example, a press button is connected to an ATTiny microcontroller pin. The button is configured as active *low* and there is a voltmeter that measures the value of the voltage at the pin. As such, when the button is pressed, 0V should be observed on the pin. However, in the example when the button is both pressed and unpressed the voltage value on the pin is the same. Fix the example such that when the button is unpressed the voltage observed at pin is 3.3V or *high*.

> *Tinkercad has a schematic view for circuits. This can be helpful at times when trying to see if you built a circuit correctly. You can access that view by clicking on the schematic icon in the upper right of the simulator page.*

> *The Tinkercad simulation environment does simulate the effect of **noise**. However, in a real-life hardware, when the button is unpressed, you might see a voltage value larger than 0V.*

👾 Drive Current: This example uses an Ammeter to measure the current going through an LED connected to a microcontroller pin. Note what the current is at the start of the simulation. Increase and decrease the value of the resistor and notice what happens to the current value. Also notice how the intensity of the LED light changes.

👾External Transistor Driver: This example shows how a load can be driven by an external transistor. Note how now there are two different power supplies or batteries. Also how the voltage values of the batteries are different. Now if you were to change the current going through the LED, which resistor value would you change. Try to replace the LED with a DC motor check how the revolution per minute (rpm) changes with changing current.

> *In tinkercad a negative rpm value indicates a counter clockwise (CCW) direction of rotation, while a positive value indicates a clockwise (CW) direction of rotation.*

👾 H-Bridge: In this example an **H-bridge** circuit is available for you to finish and control a motor. For that, you can use the ATTiny device or an Arduino blocks in Tinkercad to create a simple application with code blocks. Note that there are two lines for you to control; a yellow line and a turquoise line. The yellow line activates one of the upper transistors with one of the lower transistors to drive the motor in one direction. Conversely, the turquoise line activates the opposite two transistors to drive the motor in the opposite direction. Both lines cannot be **high** in the same time.

2.5 Questions

1. You have a microcontroller with pins that have a current drive of 30mA. Additionally, you have an external load that requires 50mA of current. Would an external *driver* be necessary?

2. You have a microcontroller with pins that have a current drive of 50mA. Additionally, you have an external load that requires 30mA of current. Would an external *driver* be necessary?

3. LED intensity is controlled by reducing the average amount of current going through it. This often done using software. If you were to control the average amount of current delivered to an external load, how would you do it?

4. If you have a board with an external *pull-up* at an input pin. Assume you inadvertently configure a microcontroller with another internal *pull-up* or *pull-down* what do you think happens in either case?

3. Interfacing Timer/Counter Peripherals

Timers and counters, as inputs, are peripherals that can be used to either measure the time/duration between external events or count how many times an event occurs. On the other hand, as outputs, timers and counters can also be used to generate different waveforms. Timers and counters commonly interface to the outer world through three modes found in microcontrollers:

1. **Input capture** Mode
2. **Output compare** Mode
3. **PWM** Mode

3.1 Input Capture

Input capture mode is a feature that enables the precise measurement of timing between external events, like signal transitions or pulses. It works by continuously monitoring the state of a designated input pin. When a specified event occurs, such as a **rising** or **falling edge** of a signal, the system captures the current value of the timer peripheral. This captured value, along with subsequent ones if needed, is then used to calculate the time difference between multiple events. This mode allows for accurate time interval measurement, frequency calculation, and event synchronization, making it useful for applications like measuring the speed of motors, generating precise time delays, or monitoring time-based processes with high precision.

Interfacing considerations for **input capture** are similar to digital inputs.

3.2 Output Compare

Output compare is a feature that allows a microcontroller to generate precise timing-based events or signals based on a comparison between a timer's value and a pre-set reference value. It works by configuring a timer to count up or down, and then when the timer's value matches or exceeds the specified reference value (the "compare" value), an action is triggered, such as toggling an output pin, generating a pulse, or initiating an interrupt. This capability is commonly used for creating accurate time delays, generating periodic waveforms, or synchronizing processes that require specific timing intervals or patterns.

Interfacing considerations for **output compare** are similar to digital outputs.

3.3 Pulse Width Modulation (PWM)

3.3.1 What is a PWM?

Pulse width modulation (**PWM**) is a special waveform generation technique that controls the amount of on time in a square wave signal. **PWM** comes from communication system origins but found a lot of use in embedded systems. A traditional square wave signal spends 50% of its time in a single period as "on" or **high** and the other 50% as "off" or **low**. Figure 16 shows a traditional square wave with a fixed frequency. Note that the period is denoted with T, the on time is denoted with T_{on}, and the off time is denoted with T_{off}.

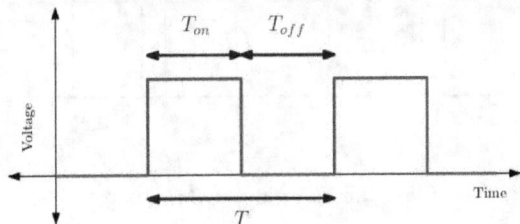

Figure 16. A Traditional Square Wave

PWM signals are essentially square waves that allow the alteration of on-time (T_{on}) in a single period of a square wave. This means that T_{on} need not be only 50% active within a period. Consequently, with **PWM**, the on-time in a square wave signal can be altered all the way from 0% (signal completely off) to 100% (signal always on). Note also that we are changing the "Pulse Width" within a period, thus the naming.

A common term used in **PWMs** is the **duty cycle** which is defined as the ratio of on-time to period of the signal. Or in other words the percentage of time that the pulse is on within a single period. Figure 17 below shows different examples of **PWM** waveforms with different **duty cycles** or on-times. Note that while the **duty cycle** of a signal changes the period (and thus frequency) remains the same.

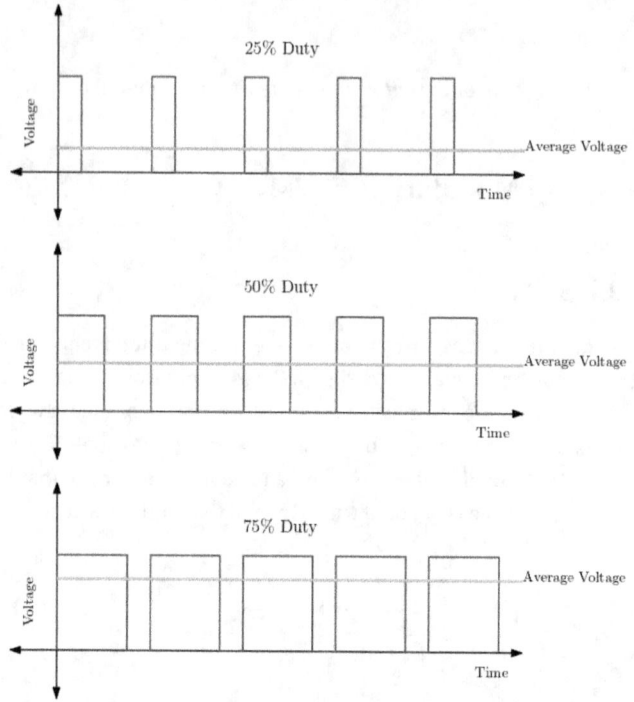

Figure 17. Various PWM Signals

Though what does the "Average Voltage" in Figure 17 mean? Read on to find out.

3.3.2 What is the Purpose of PWM?

While **PWM** comes from communication systems, they are used for a totally different purpose in embedded. Meaning that we are not using **PWM** to send any data. Instead remember how earlier it was highlighted that using digital output to control current delivered to an actuator can be cumbersome. **PWM** offers a more convenient approach through generating a waveform via hardware. **PWM** is like turning a flashlight on and off really quickly. When it's on more than it's off, it's like shining brightly. When it's off more than it's on, it's like shining dimly. In the context of current, **PWM** is used to control how much electricity flows. By turning the electricity on and off rapidly, one can control the average amount of voltage applied at the output and thus current flowing through a load. It's a bit like adjusting the faucet to let a little water out, then turning it off, then a little more water, and so on. This way, you can control the "average" amount of water flowing, or in this case, the average amount of current.

This concept was shown in Figure 17. Note when the *duty cycle* increases then the average corresponding voltage increases. The nice part is that the average voltage ratio to the full scale voltage is the *duty cycle* you need. For example, if you have a *PWM* signal that is 0V when off and 5V when on, and you want an equivalent voltage of 2.5V to appear. This means that you need a 2.5/(5-0) = 50% *duty cycle* to achieve 2.5V average output equivalent. Similarly, the other way around if you were to apply 75% duty on a 0-5V signal you would get (5-0)*0.75 = 3.75V equivalent output voltage.

It becomes obvious that any application that can benefit from current control can leverage *PWM*. Examples include controlling LED intensity, motor speed, linear actuators position, servo motor position, amount of heat...etc.

3.3.3 PWM Interfacing

Interfacing considerations for *PWMs* are more or less similar to digital outputs. The point however is that when it comes to the driver circuits introduced earlier, instead of using a regular digital output a *PWM* can be used. This allows for control of the average current delivered to the load if needed.

3.4 Simulation Examples

🕹 PWM: In this example, a servo motor is being driven by *PWM* from an Arduino UNO board. In the code block of the simulation, try changing the servo angle and rerun the simulation. See how the *PWM duty cycle* changes. Try adding a voltmeter and measure the voltage at the *PWM* output. Note also how the voltage will change as the duty cycle changes. This is the average voltage delivered to the servo. Calculate the average voltage by hand and compare, is it consistent with the reading you are getting?

🕹 **PWM Driving Loads**: Tinker the existing earlier examples of Drive Current, External Transistor Driver, and H-Bridge to control each with a *PWM* instead of GPIO output. Using *PWM* you should be able to control the motor speed and the LED intensity by increasing and decreasing the average voltage delivered.

3.5 Questions

1. What is the average voltage delivered to a load for a **PWM** output pin that has a voltage range of 0-5V and a **duty cycle** of 25%?

2. For a **PWM** output pin that has a voltage range of 0-3.3V what **duty cycle** needs to be applied to achieve an average voltage of 1V?

3. Servo motors are controlled by **PWM** signals that have a period of 20 ms. Within a single period, if the on time is 1ms then the servo motor position is -90°. If the on time is 1.5ms then the servo motor position is centered (0°). Finally, if the on time is 2ms, then the servo motor is in its 90° position. Calculate the **duty cycle** needed for each of the positions.

4. Motors are typically connected through driver circuits and their speeds controlled through the voltage applied across them. This is done through **PWM**s. Assume a circuit similar to the one in Figure 14, however instead of an LED you'd have a motor and V_{cc} is 12V. You can also ignore the effect of the resistor. Assume that you want to run the motor at 1200 rpm and for that the average voltage applied across the motor has to be 9V. What should be the **duty cycle** of the **PWM** at the output pin to achieve 9V across the motor? (*hint*: the output voltage of the GPIO pin doesn't matter, only its **duty cycle**)

4. Interfacing ADC Peripherals

Analog to Digital Converters (**ADC**s) convert continuous analog signals, such as voltage or current, into discrete digital values that can be processed by a microcontroller. This conversion is essential because many real-world signals, such as audio, temperature, or sensor data, are inherently analog in nature, though digital devices work with discrete, binary data. **ADC**s enable the seamless integration of analog and digital systems, allowing us to capture, store, and manipulate analog information in a digital format. This capability is crucial in applications ranging from audio recording and telecommunications to medical devices and industrial automation, where precision and accuracy are paramount.

There are three main parameters that govern the operation of **ADC**s:

1. Its *reference voltage* (commonly denoted as V_{ref}) which controls the range of voltages the **ADC** can read and defined in Volts.

2. Its *resolution* which controls how accurate a reading is and is defined in bits.

3. Its *sampling frequency* which controls the range of frequencies the **ADC** can read and is defined in samples/sec (Hz).

ADCs typically have two *reference voltages* a positive reference and a negative reference. The positive reference marks the maximum value range or readable values. The negative reference, on the other hand, marks the minimum value in the range of readable values. It is common for the negative reference to be tied to ground or 0V. In this case V_{ref} refers to the value of the positive reference. For example, an ADC that has a V_{ref} of 5V can measure a range of values from 0V to 5V.

ADCs measure voltages by taking *samples* at different time instances. Each *sample* is a bit value that corresponds/maps to a voltage value in an instance of time. The speed at which an **ADC** takes *samples* is referred to as the *sampling frequency* or *sampling rate*. A higher *sampling frequency* indicates that more samples are being taken in a time period. Though the question is, what should the rate be? Typically it should follow the *Nyquist criteria*. The *Nyquist criteria* states that the *sampling frequency* should be at least twice as fast as the highest frequency of the signal you are trying to sample. For example, if you are trying to sample a 20kHz sine wave, this means your *sampling frequency* should be at least 20kHz*2 or 40kHz.

The number of bits each **sample** represents is defined by the **ADC resolution**. For example, if the bit **resolution** of an **ADC** is 8-bits, then the range of values it generates is from 0 to 2^8 (0- 255). As such, one can simply think of an **ADC** as a peripheral that does a mapping from a range voltages to a range of bits. As such, if we have an **ADC** with V_{ref} of 5V and 8-bits of **resolution**, then the range of 0-5V will be mapped to the range of 0-255.

When you think about it there is a bit of a problem here. This is because the analog range of 0-5V has an infinite amount of values. As an example, I can have a voltage that is equal to 3.4567235434 V in the 0-5V range. On the other hand, the 0-255 digital (binary) range is finite, meaning that it can represent exactly 256 (255+1) values. Also each value in the digital range has an exact voltage it represents. So what does an **ADC** do in this case? There is something referred to as **quantization**. What **quantization** does is simply rounding up or down to the closest finite value. This means that a range of analog values would map to a single digital (binary) value. As a result, the less **resolution** we have, the more analog values would map to a single digital one.

This leads us to another question. Aren't we losing accuracy here? The answer is yes, but then the question becomes how much is acceptable to our application? Meaning if I want to measure voltages in the μV range but all I can get is mV range, what can I do? The answer is in increasing the bit **resolution**. There are **ADC**s out there that have resolutions of 8, 12, 14, and 16. However, increasing the **resolution** too much might introduce effects where one starts sampling unwanted signals or **noise**.

Noise can be simply thought of as any unwanted signal that couples (or gets added) on the signal you are trying to measure. There are several ways where **noise** can find its way into measurements and mess them up. **Noise** sources are numerous and there are several ways to protect against them. In what's coming up next we'll cover some of the main common approaches in interfacing electronics.

In summary, in interfacing **ADC**s, there are two measures our interfacing electronics need to focus on:

1. Reducing the effect of **noise**.
2. Achieving compatibility with the **ADC**'s input range of voltages.

4.1 Comparators

A **comparator** is an electronic circuit used to compare two input signals and determine their relative magnitudes. Its primary purpose is to produce an output signal that indicates whether one input is greater or less than the other. It can also be thought of as the analog version of an if statement in software. A **comparator** is a special configuration of an electronic component referred to as an **operational-amplifier (op-amp)**. **Op-amps** can be viewed as super-sensitive, adjustable voltage amplifiers. **Op-amps** take in electrical signals, amplify them, and produce an output voltage.

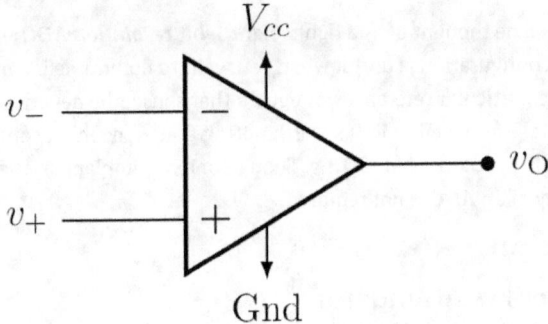

Figure 18. A Comparator Circuit

Figure 18 shows the schematic diagram of a **comparator** circuit. Here's how a basic **comparator** operates:

1. **Inputs**: A **comparator** typically has two input terminals, often labeled as the non-inverting input (+) and the inverting input (-). These inputs accept the two signals that you want to compare. The non-inverting input is often connected to some reference voltage, while the inverting input is connected to the signal being measured or compared.

2. **Output**: The *comparator* has a single output terminal (v_O) that provides the comparison result. The output is usually a binary signal, which means it can be in one of two states: high or low, representing "greater than" or "less than" respectively, or vice versa, depending on the comparator's configuration.

3. **Comparison Process**: The *comparator* constantly compares the voltages present at its two input terminals. It internally subtracts the voltage at the inverting input from the voltage at the non-inverting input.

   - If the voltage at the non-inverting input (+) is greater than the voltage at the inverting input (-), the *comparator's* output goes to its positive supply voltage level (usually denoted as "high" or logic level "1").

   - If the voltage at the inverting input (-) is greater than the voltage at the non-inverting input (+), the *comparator's* output goes to its negative supply voltage level (usually denoted as "low" or logic level "0").

A *comparator* can be thought of as a lightweight 1-bit *resolution ADC* as v_O represents a single bit. This configuration is useful when you want to determine if a measured value is under (or over) a particular reference voltage. In that sense, the negative terminal would have a fixed voltage attached to it. It's worth noting that some microcontrollers have comparators built-in additional to *ADCs*. Though one can implement their own externally if the complexity of an *ADC* is not required.

4.2 Amplifiers vs Attenuators

An *amplifier* and an *attenuator* are two fundamental electronic components or circuits used to manipulate the amplitude or strength of electrical signals, but they serve opposite purposes.

An *amplifier's* purpose is to increase the amplitude or strength of an electrical signal while preserving its shape and frequency content. In other words, it magnifies the input signal. *Amplifiers* often use components such as *transistors* or *op-amps* to achieve signal amplification. The input signal is applied to the *amplifier's* input, and the output provides a larger version of that signal. *Amplifiers* can be designed to have various gain levels, allowing for precise control over the degree of amplification.

An **attenuator's** purpose, on the other hand, is to reduce the amplitude or strength of an input electrical signal while maintaining its shape and frequency content. It is used to decrease the signal's intensity. **Attenuators** can be formed by passive devices consisting of resistive components. They work by introducing resistance into the signal path, which leads to a voltage drop and, consequently, signal **attenuation**. The degree of **attenuation** is specified in decibels (dB), and **attenuators** can be designed with various **attenuation** levels, including fixed and variable **attenuators**.

Attenuators are probably more commonly seen in interfacing **ADCs**. This is because the range of outer world voltages to be measured are much larger than the range of voltages an **ADC** is compatible with. Take for example an automotive embedded device meant to measure a car battery. Ranges of V_{ref} typically do not exceed 5V in many devices, however, a car battery voltage can reach up to 16V. In this case the range of the car battery voltage needs to be **attenuated** to match the range of the **ADC**.

Finally, it is worth mentioning that there are a lot of different forms which **attenuator** and **amplifier** circuits can come in and are not within this book's scope. Nevertheless, in the next section one very common type of **attenuator** circuit used in embedded applications is going to be introduced.

4.3 The Voltage Divider

The **voltage divider** is a specific type of **attenuator** circuit that is relatively simple to build. **Voltage dividers** are fundamental in embedded electronics for controlling and scaling voltage levels. It is also common to see a **voltage divider** preceding an **ADC** pin. A **voltage divider** circuit is a simple electrical circuit consisting of two resistors connected in series as shown in Figure 19.

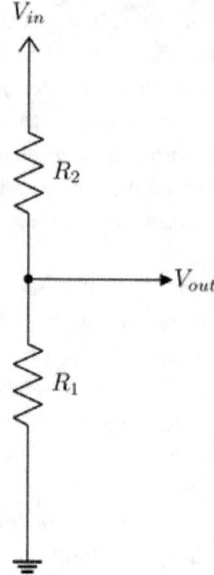

Figure 19. A Voltage Divider Circuit

A **voltage divider** is used to divide an input voltage into a smaller output voltage, proportionate to the resistor values. The voltage at the output is determined by the ratio of the second resistor's resistance to the sum of the two resistors' resistances. This relationship is described by the following **voltage divider** formula:

$$V_{out} = V_{in} * \frac{R_1}{R_1 + R_2} \tag{1}$$

A good general rule of thumb is that when you see $R_1 = R_2$ that means that V_{in} is divided by half. So for example if the range of V_{in} is 0-5V, it scales to 0-2.5V at V_{out}.

Voltage dividers have various applications, such as providing reference voltages, biasing transistors, or scaling analog signals. Most commonly, they are seen in stepping down or **attenuating** analog signals read by **ADCs**. **ADC** pin voltages have to always fall within the reference voltage range. If, for example, an **ADC** has a positive reference voltage of 5V and negative of 0V, then all measured signals have to fall within that range. Let's say for example, for the 5V reference **ADC**, you want to measure a signal that goes up to 10V. In this case you'd have to design a **voltage divider** that scales down the signal by half at least. As such, a divider with $R_1 = R_2$ does the job. Obviously you have to compensate for effect of the **voltage divider** in your software to recover the original value pre-division.

4.4 Single-ended vs Differential Measurement

ADCs sometimes have more than one option for measurement. Most commonly, you would see some pins on a controller that are marked as an analog pins. Those are pins that can be configured to do voltage measurements using some internal *ADC*. As such, you probably also would have a device like a sensor that has an output voltage terminal that you connect to one of those pins. Subsequently, you would measure the voltage on that pin. However, what is strange about that is that voltage is always measured between two points (for example, if you were to use a voltmeter). That's why it's called *voltage/potential difference* because it represents the difference in voltage between two points. Though the question that begs itself is if that's the case, then how is a microcontroller *ADC* able to measure voltage using a single pin?! The answer is because the other reference point is ground which comes from the *ADC* itself. This is referred to as a *single-ended measurement* because, well, a "single" wire is used for measurement. What is crucial in this type of measurement though is to ensure that the ground among the measuring and measured devices is common. Otherwise, measurements might not be accurate.

A less common approach, but existing on some *ADCs*, is the option of using two pins/wires for measurement. This is referred to as a *differential measurement*. In this case, the reference is not ground but rather another non-zero voltage. What an *ADC* does in this case is measure the voltage difference between the two wires/pins. In fact, there are devices that exist in several application areas like medical, industrial, and automotive, among others, that provide *differential* output.

Though the question now is why are there two types of measurement? The answer is reliability and accuracy. *Differential measurement* is more immune to *noise* than *single-ended*. However, *differential* signals require an additional wire so can be considered more costly and less optimized than *single-ended*. Generally, many applications don't require the level of *noise* immunity that *differential* signals provide. Additionally, taking the appropriate system measures and following good design guidelines can really enhance the quality of *single-ended* measurements. *Differential* signals are also commonly seen in signaling or wired communication systems. These will be discuss in a later portion of this book.

4.5 Filters

Filters, as the name implies, are electronic circuits that **filter** or keep out unwanted signals, otherwise known as **noise**. **Filter** circuits can be either **active** or **passive**. **Active filter** circuits typically have better properties but are more costly to implement, thus less commonly seen in embedded applications. **Active filters** are called as such since they use **active** electronic components like **op-amps** in their implementation. **Passive filters** on the other hand, are referred to as such since they are composed of **passive** components only like resistors, inductors, and capacitors. **Passive filters** are cheaper to implement, and more commonly seen in embedded applications.

There are three types of filters:

- **Low Pass Filters**: These **filters** allow signals with low frequencies to pass and filter out higher frequencies.
- **High Pass Filters**: These **filters** allow signals with high frequencies to pass and filter out lower frequencies.
- **Band Pass Filters**: These **filters** allow a certain range of frequencies to pass and filter out anything above or below that range.

ADC pins commonly measure DC voltages that are low frequency in nature. As such, we don't want any high-frequency **noise** to mess them up. As a result, the solution comes in using a **low pass filter**. **Low pass filters** are commonly seen preceding **ADC** pins and come in the form of a resistor paired with a capacitor as shown in Figure 20.

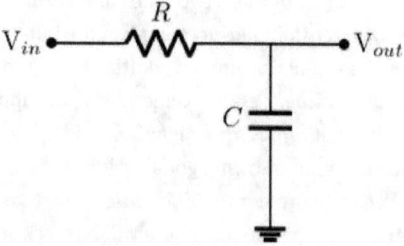

Figure 20. A Low Pass Filter Circuit

So how does this work? All you need to know is that capacitors are known to form an **open circuit** when encountering low-frequency signals. Conversely, capacitors form a **closed circuit** when encountering high-frequency signals. The resistor, on the other hand, exists to control up to which frequencies should the **filter** allow to pass.

It is worth mentioning that a **low pass filter** is also commonly known as an averaging circuit since it "averages" or smoothes the signal that passes through it. A **low pass filter** can also be implemented in software by averaging a set of samples taken consecutively by an **ADC**. Sometimes though, the overhead in software does not necessarily justify the purpose if its simple enough to implement in hardware.

Sometimes you might see a capacitor in the middle of a schematic dangling off of a power supply line to ground or in parallel to some other load. This is referred to as a **bypass** or **decoupling** capacitor. Although the names are often used interchangeably there is a slight difference in the purpose. However, both **bypass** and **decoupling** capacitors are also **low pass filters** and can be thought of as tiny electronic helpers that keep the electricity supply stable and clean. They act as sponge-like **filters**, soaking up fast and noisy electricity changes, making sure devices get steady and quiet power. They can be thought of as electrical shock absorbers, reducing jitters and glitches in signals. **Decoupling** capacitors are also commonly seen placed close to pins on many ICs.

Conversely, in audio applications, low frequency DC signals can cause an adverse effect on speakers. As such, a **high pass filter** is required to reject low frequency **noise**. This comes in the form of an in-line capacitor also commonly known as a **DC blocking** capacitor. An example of such a circuit is shown in Figure 21.

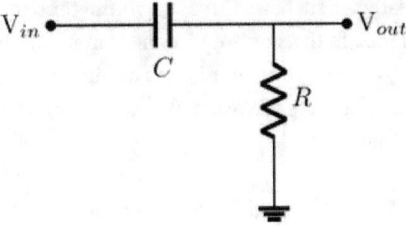

Figure 21. A High Pass Filter Circuit

> **ADC** *measurement pins are commonly preceded by a low pass filter to eliminate high frequency noise from sources like power lines. These filters can introduce a problem in sampling if the sampling exceeds a certain rate. As such, the **sampling rate** should be high enough to meet the **Nyquist criteria** but also less than frequency that the preceding filters are eliminating.*

4.6 Simulation Examples

Comparator: In this example, a standard *op-amp* known as the 741 is used in a *comparator* configuration. The voltages of two power supplies are being compared to each other and the resulting output monitored with a voltmeter. The *op-amp* itself is powered with a 9V battery. Therefore its output it expected to swing between 9V and ground (0V). Try to alter the input power supply values and monitor how the output changes. Also, the circuit in this form is actually not compatible with with 5V or 3.3V microcontroller pins. Using 9V might have an adverse effect on microcontroller input pins. How would you change this circuit to make it compatible with 5V and 3.3V controllers?

Voltage Divider: This example is a simple voltage divider. Experiment with different values of R_1 and R_2 to achieve different ratios. Monitor how the output voltage is changing.

Single Ended vs. Differential Measurement: This simulation is meant to give you a sense on the difference between **single ended** and **differential** measurements with and without **noise**. An **ADC** can actually be thought of as a voltmeter that keeps taking measurements periodically. As such, we can use a voltmeter to demonstrate a comparison between the measurements. In the single ended measurement, there is only one voltage source (or power supply) and the voltmeter is measuring the voltage resulting from it. In the differential measurement there are two power sources and the voltmeter is measuring the difference between them.

In the circuits without **noise**, try changing the power supply voltages. What do you notice as a difference in measurement? Also to give you a feel of why differential measurements are more robust, **noise** in the form of offset is added. This simulates the effect where some voltage gets coupled on the measurement lines. Notice how is the **single ended** measurement affected compared to the differential measurement?

📌 Filters: This example simulates a **low pass filter** and a **high pass filter**, however does not specify either one. The simulation shows that there are two output waveforms from two circuits that share the same input signal. Examine the circuits and the schematic, try figure out which one is the **low pass filter** and which is the **high pass filter**. Also experiment with the changing input frequency and see what happens to the output.

What should happen is that for the **low pass filter** when exceeding a certain frequency the output signal gets **attenuated**. Otherwise, the output signal should be the same as the input signal. This behavior is the opposite for a **high pass filter**. You'd have to keep an eye on the voltage value on the y-axis to identify any attenuation. Also find out what is the frequency at which the output starts changing for either filter. The filters are designed such that the frequency is similar.

> The frequency boundary at which above or below it the signal is attenuated is referred to as the **cutoff frequency**. For a **high pass filter** any signal frequency below the cutoff frequency gets attenuated. Whereas in a **low pass filter**, any signal below the cutoff frequency is attenuated. For passive filters there are several tools out there that allow you to calculate the values of the passive components needed for a desired **cutoff frequency**. The **digikey low-pass filter/high pass filter calculator** is one example.

4.7 Questions

1. You are trying to measure a signal that has a frequency of 30Hz. What should be your minimum **sampling frequency**?

2. You are trying to measure a signal that has a frequency of 10Hz. However, the ADC measurement pin has an external **low pass filter** that removes any frequencies above 50Hz. What is the range of **sampling frequencies** you can apply?

3. You have an **ADC** with V_{ref} of 3.3V. You are trying to measure the voltage across a 9V battery. If $R_1 = 1k\Omega$, calculate the value of R_2 you'd need to make the battery voltage compatible with the **ADC** measurement capability.

4. You have a signal input that has a range 0-12V. You want to detect if the voltage drops below 5V. How can you use a **comparator** for that? How would you connect to each of the terminals?

5. Consider an **ADC** doing a **single-ended** measurement with a voltage of 1.8V at the **ADC** input pin. The measurement would be straight forward and correspond to the mapped decimal/digital value of 1.8V at the pin (dependent the **resolution** and V_{ref}). In a **differential** measurement, there are two pins (a + and -) similar to a **comparator**. Say the + pin had a voltage of 2V and the - pin had a voltage of 1.5V, what voltage do you think the resulting mapped decimal/digital value would correspond to?

6. Consider the same example in the previous question. Say some **noise** coupled on the lines being measured. We can for simplicity sake consider this **noise** as a constant voltage of 0.5V added to each line being measured. So for the **single-ended** measurement pin the voltage now is 1.8V+0.5V=2.3V. Also for the **differential** measurement pins its now 2V+0.5V=2.5V and 1.5V+0.5V=2.0V. How would the resulting mapped decimal/digital value change in each case?

7. What type of **filter** would you use if you want to eliminate any **noise** above 20kHz?

8. What type of **filter** would you use if you want to eliminate any **noise** below 20kHz?

9. What type of **filter** would you use if you want to eliminate any **noise** above 20kHz and below 10kHz?

5. Interfacing Serial Communication Peripherals

The world of communication system electronics is vast. Nevertheless, whether wired or wireless, some main goals communication systems try to achieve include:

- Long communication distance
- High data rate
- Minimum amount of errors (high immunity to *noise*)
- Low power consumption
- Low resource consumption (Ex. pins or components)
- Number of entities that can be addressed

Obviously, this not all achievable, so different systems (or protocols) exist to accommodate different application needs. This section focuses on wired serial communication interfaces which are common in embedded systems. The most common protocols being *UART*, *SPI* and *I2C*. Obviously there are many others like CAN, USB, and Ethernet, among others. However, the focus of this book is not how these protocols work but rather to identify the interfacing circuits. In that context, there is a lot of commonality in the ideas behind how the interface electronics look like. You'll notice also that a lot of the concepts carry over from what we have already covered in this book.

SPI and *I2C* are used commonly for on-board communications with devices like sensors, screens, or memory devices, among others. Both are multi-point in the sense that multiple entities can be addressed. The difference of using one versus another depends on bandwidth and on-board resources. *SPI* for example offers much higher bandwidth than *I2C* but utilizes more wires. *SPI* is a 4-wire interface whilst *I2C* is a two-wire interface. *I2C* is sometimes even referred to as such, a *two-wire interface* (or TWI). Both *SPI* and *I2C* are also *synchronous* which means that they have a have a dedicated clock line accompanying the data line(s). At a pin level, *I2C* uses the *open-drain* configuration, whilst SPI uses the *push-pull* configuration.

UART is a point-to-point protocol in its simple form. Meaning that only one entity can talk to only one other. *UART* uses a single wire (in ***push-pull*** configuration) to transmit (an a separate wire to receive as well) and is an interface typically used to program a controller or log monitoring messages. Additionally, *UART* is ***asynchronous*** meaning that it does not have a dedicated clock line. Synchronization is achieved instead by introducing data overhead.

5.1 Differential Signaling

Pins on a controller have low voltages and are ***single ended*** in nature. Moreover, if using ***asynchronous*** communication, which is self-synchronizing, results in a lot of sensitivity to ***noise*** if traversing long distances. As a result, one option is to introduce ***differential signaling***. This is a similar to the idea introduced in the *ADC* measurement earlier section. Here the voltage difference between two wires is measured instead of measure the voltage on one to increase ***noise*** immunity.

In order to achieve this, a challenge is that some serial communication pins on a controller are not ***differential*** by nature. As such, the pins/lines need to be fed to a ***transceiver*** IC that converts them from ***single-ended*** to ***differential*** signals. One of the common transceivers used with *UART* is a RS-485 transceiver. RS-485 is a signaling standard that provides robust transmission using ***differential signaling*** to minimize ***noise*** interference.

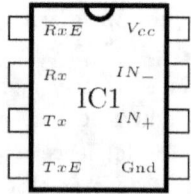

Figure 22. Example of a RS-485 Transciever Pinout

RS-485 transceivers come in different forms. However, one pinout of a RS-485 transceiver might look similar to the on in Figure 22. Note on the right side the IN+ and IN- pins. These are the pair of ***differential*** lines that would connect to the transceiver. On the left side, there is the Tx and Rx pins. Each of those pins is ***single ended*** and propagate the data that is either being transmitted or received. The RxE and TxE pins are only enable pins. The MAX22500E IC is a type of RS-485 transceiver.

Recall that each pair of *differential* lines would convert to one *single ended* line. Though something looks weird in Figure 22, we have one pair of *differential* lines corresponding to two separate *single ended* lines. This is because this transceiver IC supports *half-duplex* communication. This means that we can either transmit or receive but cannot do both simultaneously. This also means that the *differential* lines would be either used to transmit or receive but cannot do both at once.

5.2 Voltage Level shifting

There is another way to address the *noise* traversing long distances on *single ended* wires; by voltage level shifting. Remember the issue highlighted earlier was that the pin voltages levels are not that high. Also long lines can have higher resistances, dropping voltage levels even further resulting in increased *noise* susceptibility. As such, another approach could be increasing the voltage over the communication wire and then drop it down again at the receiver. A signaling technique commonly used to achieve that, also typically used with *UART*, is RS-232. Similar to the RS-485, there would be an external IC that converts to/from controller pin voltage levels to RS-232 levels. The MAX232 IC is a type of RS-232 transceiver.

5.3 Bridging

UART is the protocol commonly used to interact with a host PC in embedded. This could be for logging or downloading code to the microcontroller. Older computers used to have what was called DB9 connectors (shown in Figure 23). These connectors actually had pins that supported RS-232 signaling. However, PCs nowadays do not have *UART* compatible interfaces, but rather interfaces like USB. Additionally, we connect development boards to a host PC using USB interfaces. Does that mean that the *UART* protocol is not being used? The answer is no, and its because an *interface* or *bridge* IC is being used. A *bridge* device is one the converts from one signalling protocol to another. One famous device is the FTDI chip used for programming microcontrollers. Also the current Arduino UNO for example uses another microcontroller with a dedicated program to do the USB to *UART* conversion. It's the IC that gets placed close to the USB socket on an Arduino UNO board. This device comes with preprogrammed firmware and is the interface commonly used to support debugging as well.

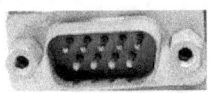

Figure 23. A DB9 Connector

6. Voltage Regulators

Embedded systems are often powered by a variety of different power sources with a wide range of voltages. For example in a car, electronic systems are powered by the car battery that has a voltage ranging from 10 to 16V. On the other hand, a maker board could be powered by a 9V battery. The challenge is that microcontrollers commonly require a stable power source of either 3.3V or 5V. One might think, why not just use an *attenuator* type circuit like a *voltage divider*? The reason is that a *voltage divider* would step down the voltage to a certain range, though it cannot guarantee the output voltage stay stable.

The answer comes in *voltage regulators*. While not necessarily part of a peripheral interface, *voltage regulators* exist much too often in embedded systems to ignore. As the name implies, a *voltage regulator* "regulates" voltage. *Voltage regulators* take a wide range of voltages as input from different types of power sources, to provide a constant voltage as output. This constant voltage output would be one compatible with a microcontroller system like 3.3V or 5V.

Given the above, *voltage regulators* form a core component of an embedded electronic unit. This is because it becomes one of the main sources of power. As such here are a few things to be mindful of:

- **Voltage regulators have an output current rating**. All circuitry in an electronic unit draw current, including the microcontroller. However, though much higher than a GPIO pin, a *voltage regulator* does have a limit of how much current it can source. The total current draw of components in an electronic unit cannot exceed that limit or the regulator might overheat and get damaged.

- **It's is common to see a capacitor connected to ground at voltage regulator outputs**. This capacitor serves as a *filter* like explained in the earlier sections and smoothes the voltage even further by removing any high frequency *noise*.

- **ADC measurements are dependent on having a reference voltages**. Without a stable source, the measurement won't be accurate. As such, the *ADC* reference voltage value is typically driven from the regulator.

- **Controllers (and their pins) are rated for a certain voltages like 5V or 3.3V**. Exceeding those ratings could cause damage to the controller and/or its pins. As a result, *pull-up*s are driven by the output of the same voltage regulator powering the microcontroller.

Actuators like motors do not require regulated voltage. As a result, their driving circuitry is not typically supplied power by the regulator. A common rule of thumb is that all circuits that require stable digital voltage levels (Ex. 0V for digital low and 5V for digital high) are ones that should be powered by the voltage regulator. Moreover, certain actuators like motors might have current requirements that go beyond what a regulator can source.

6.1 Simulation Examples

Voltage Regulators: This example should give you a sense of how voltage regulators work. There are two regulators, a 3.3V regulator and a 5V regulator, and both their inputs are being fed by the same power supply. Change the voltage on the input supply and notice what happens to the outputs as the input volage increases and decreases.

7. Oscillators

Microcontrollers require not only power to run, but also an accurate clock source. All internal circuitry in a microcontroller is synchronized to a global clock signal that comes in the form of a square wave. The clock signal in turn is generated by an *oscillator* circuit. *Oscillator* circuits are like tiny electronic "heartbeat" devices in many electronic devices. They generate continuous electrical signals, usually in the form of waves, which can be either sine waves or square waves.

An *oscillator* is similar to a swing that if pushed at the right time and with the right force, it keeps moving back and forth by itself (or *oscillating*). *Oscillator* circuits consist of electronic components, like resistors, capacitors, and transistors, arranged in a specific way. These components interact, causing the circuit to continuously switch between voltage states, creating a repeating wave pattern. This repetitive pattern happens because the components are designed to feed a portion of their output back into their input, creating a loop. This loop keeps the circuit *oscillating*, producing a continuous wave.

A special form of *oscillator* circuits commonly utilized in digital systems is a *crystal oscillator* circuit. At the heart of a *crystal oscillator* is a tiny crystal resonator made of quartz. This is because crystals provide a stable and precise clock signal. This quartz crystal has a property called piezoelectricity, which means it vibrates when an electrical voltage is applied to it. These vibrations are extremely stable and consistent.

Every quartz crystal has a natural frequency of vibration. The natural frequency at which the quartz crystal vibrates is determined during its manufacturing process. Similar to other *oscillator* circuits, *crystal oscillators* have specific arrangements to ensure they keep on oscillating. In microcontroller applications part of this circuitry is integrated internal to a microcontroller. Typically microcontrollers *oscillator* pins labelled OSC1 and OSC2 or XTAL1 and XTAL2 marking where the external *crystal oscillator* circuit should be connected.

Figure 24 shows a common configuration you would find in microcontroller circuits. One, or both resistors, not necessarily be seen in all configurations. Though typically the most important information we would need to extract from this type of circuit is the crystal frequency. This is because some tools ask about the external crystal frequency to generate internal microcontroller clock configurations.

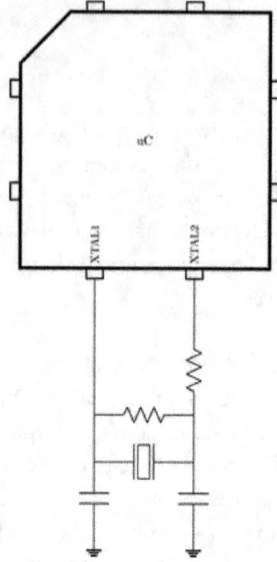

Figure 24. A Microcontroller Crystal Oscillator Circuit

8. Schematic Example

8.1 The Arduino UNO Schematic

In this chapter, we are going to look at a real life schematic and identify its circuits based what we covered. We are going to examine the infamous Arduino Uno schematic that you can find publicly on the Arduino website (https://www.arduino.cc/en/uploads/Main/arduino-uno-schematic.pdf).

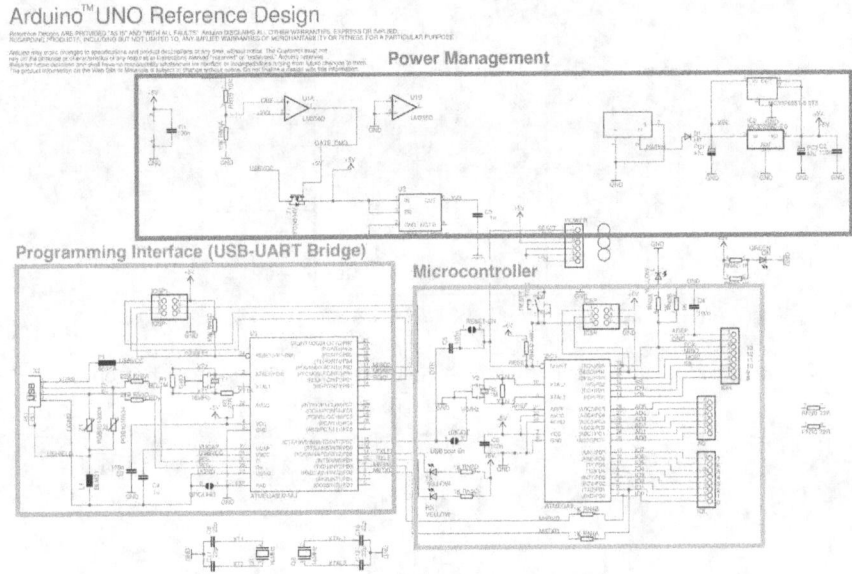

Figure 25. Schematic of the Arduino UNO Board

Before digging into details, let's have a global view of the schematic. Figure 25 highlights three main areas in the Arduino UNO schematic; the programming interface, the microcontroller, and power management. The programming interface incorporates the device used to program the UNO microcontroller. Devices that appear in symbols similar to the shown in Figure 26 refer to integrated circuits. On the other hand, symbols that look similar to what is shown in Figure 27 refer to connectors. Note how each pin also has a name and a number. Additionally, if the component has a part name, it appears below it (Ex. ATMEGA8). This is useful when we want to look up data sheets for the component. As we progress in this chapter, we will highlight different circuits from the schematic and show it relates to what we've covered in this book.

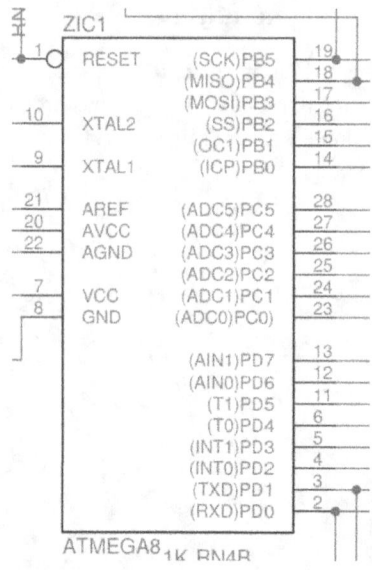

Figure 26. Symbol of an integrated circuit (the ATMEGA8 microcontroller)

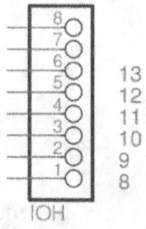

Figure 27. Symbol of an On-board Connector

The microcontroller on the UNO (denoted ZIC1 on schematic) is both programmed over **UART** and also uses **UART** for logging. As such, the programming interface also acts as a USB to **UART** bridge. On the left of the schematic you might note the USB text in a connector symbol which actually highlights the USB connector on the board. USB uses *differential signaling*. As we already know, *differential signals* come in pairs. These are pins 2 and 3 shown in Figure 28 and connect to the D- and D+ wires.

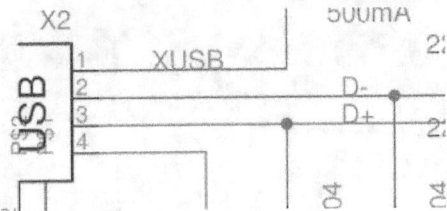

Figure 28. USB Connector Differential Lines

If you follow the D- and D+ lines, you'll notice that they connect to pins 29 and 30 of the programming controller (named U3). On the other side of the programming controller, take a look at pins 8 and 9 (M8TXD and M8RXD). These lines are connected to pins 2 and 3 of the microcontroller and also pins 0 an 1 of the Arduino connector. These are the microcontroller *single-ended UART* transmit and receive pins (note the naming for TXD and RXD for pins 2 and 3).

> *A common issue known that Arduino UNO users run into is that programming fails when board pins 1 and 0 are connected externally. Pins 1 and 0 are marked as **UART** pins on an Arduino UNO and rightly so makers use them as such. However, the issue happens because pins 1 and 0 share the same **UART** lines for programming as we've seen. Having an external connection would affect the voltages on these lines and thus cause the communication with the device to fail.*

Figure 30 is a snapshot of the upper right circuit in the power management block. This is the 5.0V regulation (or power generation) circuit for the Arduino board. Note that there are two main parts; the MC33269ST-5.0T3 and the MC33269D-5.0, if you look up the data sheets you'll find that both are *voltage regulators* that supply a stable 5.0V output. On the schematic you'll also notice that both also share the same input and the same output. Then why two? In reality on the Arduino physical board only one is used. This is a technique utilized in schematics to provide an alternative option, however it doesn't mean that both are used simultaneously (or populated).

The block on the left of Figure 30 is the on-board power jack (or barrel connector). The PWRIN signal, observed on the line from the left, is the power from the on-board power jack. This is a voltage typically is somewhere between 7 and 12V. The jack voltage is renamed from PWRIN to VIN on the wire after the diode D1 if you notice. Note also the capacitors dangling off the power lines, both at the input and the output. These are what we referred to as **bypass** and/or **decoupling** capacitors that act as **low pass filters** omitting different forms of high frequency **noise**.

> *Regarding diode D1, while we didn't cover diodes in this form, diodes are components that allow current flow only in one direction. D1 here is meant for protection purposes.*

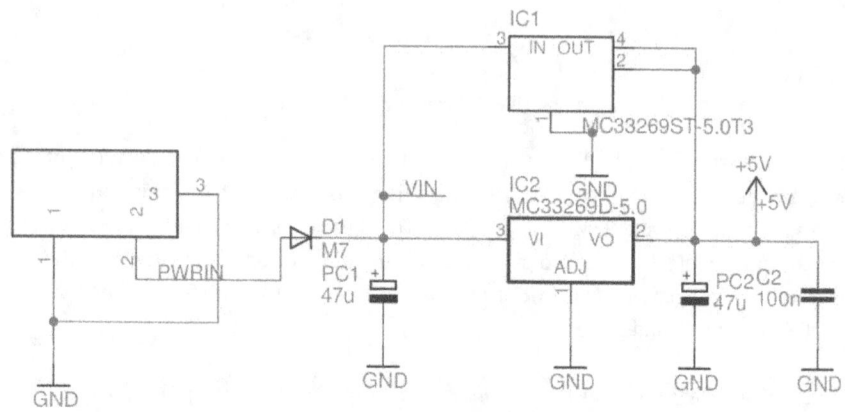

Figure 30. Arduino UNO Board 5V Regulator Circuit

Figure 31 is a snapshot of the upper left circuit in the power management block. This is the 3.3V regulation circuit for the Arduino board. Note that the first part from the left there is a **voltage divider** with VIN (same one we saw earlier) as a voltage input and 10kOhms for both R1 and R2. As we've learned, this means that the **voltage divider** is dividing the voltage VIN by 2.

The divided output voltage of the **voltage divider** is being fed into the positive terminal of a **comparator**. Also the negative terminal of the **comparator** is being fed with 3.3V. This means that the **comparator** is comparing VIN/2 with 3.3V. As such, if VIN/2 is higher than 3.3V the comparator will show a 5V output (on the GATE_CMD wire), otherwise it will show a 0V output. Why is it 5V? this is because the **comparator** circuit is powered with 5V. It is shown in the small circuit to the left of the **voltage divider**.

The output of the **comparator** is connected to a type of FET transistor. This is a transistor that is operating as a switch. One end is connected to the U2 **voltage regulator** input (the LP2985 part) and the other end is connected to the USBVCC wire (power from the USB connection). If you look up the LP2985-33DBVR part data sheet, it's a **voltage regulator** that generates a fixed 3.3V output. In essence, what this circuit is doing is determining if the value of VIN is high enough (larger than 3.3*2= 6.6V) to generate the 3.3V regulated voltage. If VIN is not high enough then the circuit switches over to USB to power the 3.3V supply on the board. The board will also resort to USB for regulated 5V power.

In summary, the Arduino UNO allows a user to supply power from different sources; either USB or through the power jack. If both are connected, the board defers to the power jack supply if it's high enough. If it's not then it switches over to powering from USB.

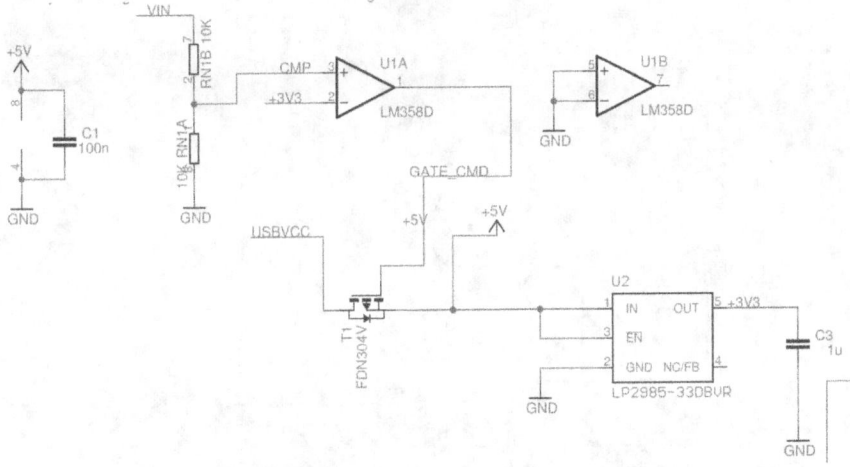

Figure 31. Arduino UNO Board 3.3V Regulator Circuit

*You might note something that looks a bit strange in Figure 31. There is an **op-amp** (U1B) with both input terminals connected to ground and the output terminal open. This is because some schematic editors require you choose a part and its associated package. If you look up the LM358D data sheet you'll notice that it comes in a specific type of package (enclosing). These packages can include several instances of the part though we only need one. As such, the extra instances will appear in the schematic and we need to sometimes connect them in a way to minimize power consumption. Leaving inputs floating is not a good choice typically. You might also find manufacturer recommendations in component data sheets for how to deal with unused instances.*

There is a similar case in the lower right part of the schematic. You'll note two resistors just hanging there doing nothing, RN3B and RN3C. Yes, resistors do sometimes also come in packages commonly referred to as resistor arrays. These are two resistors that are part of an array selected and aren't needed but appear in the schematic anyway.

Now let's take a look around the microcontroller. If you look around to the right side you should see a circuit that looks familiar shown in Figure 32. There is an LED being driven by a microcontroller GPIO pin. This is the circuit connected to microcontroller pin 19 (pin 13 on the Arduino outer connector). This is the on-board LED used for Blinky examples! In this configuration you can see that the LED turns on when the output is high.

You might have noted the two resistors each with 1 kOhm in value. These are place in a parallel configuration. In this configuration they generate an equivalent resistance of 500 Ohms.

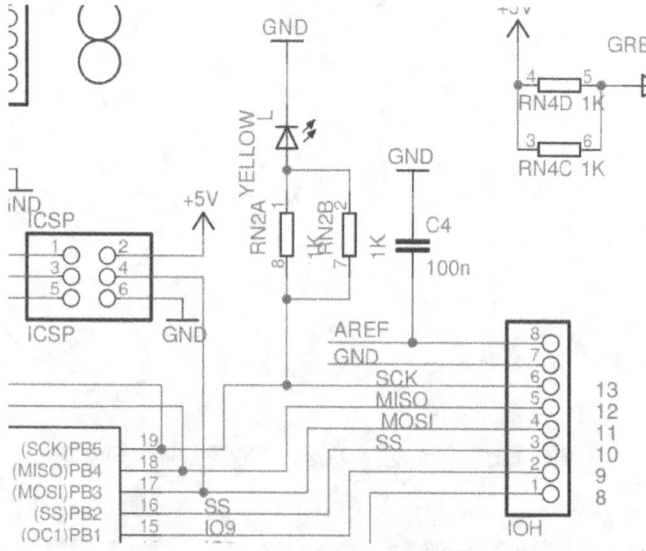

Figure 32. Main Microcontroller Blinky LED

If you look over to the programming block, you'll see a similar circuits but with the alternative **driver** configuration we mentioned. The snapshot is shown in Figure 33. These are LED circuits being driven by the TXLED and RXLED lines on the programming controller. These are the LEDs utilized to indicate that programming communication is occurring.

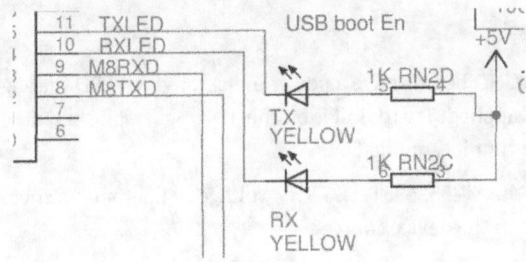

Figure 33. Arduino UNO Programming Controller LEDs

In both the microcontroller programmer blocks you should also notice the crystal oscillators. Figure 34 captures the oscillator circuit for the microcontroller. Additionally, Figure 35 captures the oscillator for the programmer.

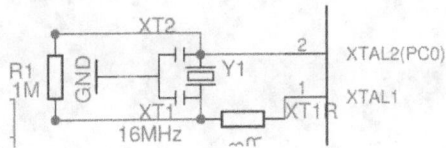

Figure 34. Crystal Circuit of the programming mircocontoller

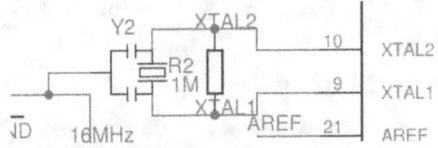

Figure 35. Crystal Circuit of the main mircocontoller

At this point we more or less covered all the electronic circuits on an Arduino UNO board! Be mindful that many development boards out there do not have circuits much more complicated than this. You also might notice that there are components scattered around that we didn't really mention. These additional components are added for protection or reduction of other noise effects typically.

8.2 Questions

1. Arduino UNO boards have a power indicator green LED that lights up when power is applied. Try to find it on the UNO schematic and determine which power output its supplied by.

2. Look up the MC33269ST-5.0T3 5V Voltage regulator data sheet. How much current can the device source?

3. Look up the LP2985-33DBVR 3.3V Voltage regulator data sheet. How much current can the device source? Is there a difference with the 5V supply?

4. If you need to control a device that needs more current than what either of the voltage regulators can support, what do you think could be an alternative solution?

9. Conclusion

Navigating the vast world of electronics may seem daunting at first, especially for those venturing into embedded systems. However, fear not, for electronics is your ally in this exciting journey. As a beginner in embedded systems, you don't need to become a master of electronics design; your primary focus is to understand their purpose and how they interact with microcontrollers.

In this ebook, we've laid the foundation for your embedded systems adventure by providing you with an essential subset of electronics knowledge. You're now equipped with the tools to configure controller peripherals effectively and program them to perform amazing tasks.

Remember, microcontrollers are the bridge between the digital and analog worlds, and they communicate with the outside through various peripherals, each with its unique way of interfacing. This book has provided common electronic terms and circuits that you'll encounter in these interfaces.

My hope is that you've found this ebook to be a valuable resource in your journey into embedded systems. Armed with this newfound knowledge, you're well on your way to creating innovative projects, solving real-world problems, and diving deeper into the fascinating realm of embedded electronics.

So proceed with confidence, experiment, and explore the endless possibilities that await you in the world of embedded systems. I believe in your potential to make a significant impact in this ever-evolving field.

Thank you for joining me on this educational journey. Your curiosity and determination are the keys to your success in the world of embedded electronics. Happy tinkering and coding!

Appendix

Links to Simulation Examples

- AC vs DC: https://tinyurl.com/44aha5cb
- Floating Pins: https://tinyurl.com/mtdfh5ut
- Driving LED with External Transistor: https://tinyurl.com/4yxu6f6r
- Drive Current: https://tinyurl.com/ym3n9z5y
- H-Bridge: https://tinyurl.com/dfd565a3
- PWM: https://tinyurl.com/2fh9pt7d
- Comparator: https://tinyurl.com/hjt7hbnf
- Voltage Divider: https://tinyurl.com/56z2y6dc
- Single Ended vs Differential: https://tinyurl.com/2s3u92cz
- Filters: https://tinyurl.com/4jnd96xd
- Voltage Regulator: https://tinyurl.com/5n86dp27